Cosmic Archeology

Cosmic Archeology

◆

Knowing the Universe
by Looking at Ourselves

Ian S. Beardsley

iUniverse, Inc.
New York Lincoln Shanghai

Cosmic Archeology
Knowing the Universe by Looking at Ourselves

iUniverse books may be ordered through booksellers or by contacting:

iUniverse
2021 Pine Lake Road, Suite 100
Lincoln, NE 68512
www.iuniverse.com
1-800-Authors (1-800-288-4677)

Because of the dynamic nature of the Internet, any Web addresses or links contained in this book may have changed since publication and may no longer be valid.

The views expressed in this work are solely those of the author and do not necessarily reflect the views of the publisher, and the publisher hereby disclaims any responsibility for them.

ISBN: 978-0-595-50477-0 (pbk)

ISBN: 978-0-595-61528-5 (ebk)

Printed in the United States of America

Contents

1

Cosmic Archeology

Janov Pelorat was a cosmic archeologist in Foundation's Edge, and in Foundation and Earth, by Isaac Asimov, searching for Earth. Humanity spread throughout the Galaxy, such an archeology could become very realistic. I have found the ratio 9/5 in the most humanly held sacred aspects of nature, gold, silver, sun, moon, water, air, ... and in other things like the most massive planets in the solar system, Jupiter and Saturn, pointing to the earth where that ratio is concerned, and in the human body temperature. I am also showing Arthur C. Clarke to be tuned into something deeply meaningful. Back to cosmic archeology, here is my bit of cosmic archeology:

http://my.opera.com/eanbardsley/blog/index.dml/tag/tarot

Also this post has elements of cosmic archeology:

http://my.opera.com/eanbardsley/blog/index.dml/tag/monolith

My basis of cosmic archeology:

http://my.opera.com/eanbardsley/blog/index.dml/tag/enigma

TAROT

Raphael, at the JCF forum has pursued four as significant cosmic value, I have pursued nine-fifths. While I have worked with numbers, he has worked with myth and its symbols. In an effort to generate 4 from the

1

golden ratio and 9/5 in connection with the earth and the 22 card cycle of the tarot we begin by finding the equation whose solution is the golden ratio. The golden ratio, or phi, as it is called, is the ratio such that the whole to the greater part is the same as the greater part to the lesser. That is a/b must be the same as b/c if a=b+c. Thus we have the two equations:

1. a/b=b/c
2. a=b+c

From 1 we have: ac=b^2
From 2 we have: c=a-b

These two yield: a(a-b)=b^2

Which can be written: a^2-ab=b^2
Or, a^2-ab-b^2=0

If we divide the last through by b^2, we get: (a/b)^2-(a/b)-1=0

This last is a quadratic in a/b. a/b is the golden ratio and can be found by completing the square. Letting a/b=x, our equation becomes:

3. x^2-x-1=0

We will not solve equation 3 for the golden ratio but will proceed to consider 9/5.

Raphael has presented the sequence: 5,14,23,32,41,50,59 ...

where we begin with five and add nine to each successive term. He has noted that the sum of the digits of each term is five. Thus this sequence embodies the principle of (9/5) a ratio I have found to exist throughout nature from atoms of gold and silver, to the moon and the sun, to water, and air, to the human body temperature and freezing temperature of water, to the very structure of the solar system itself. The above sequence is an arithmetic sequence, the nth term of which is predicted by:

4. $a_n = 5 + 9(n-1)$
5. $a_n = 9n - 4$ (equivalently)

Since the earth is the third planet, then n=3 yields:

$9(3) - 4 = 27 - 4 = 23$

As it so happens, 23 is the 9th prime number, and represents the earth. We write, from 5:

6. $23 = 9x - 4$
7. $27 = 9x$
8. $9x - 27 = 0$

We equate equation 8 with equation 3:

$9x - 27 = x^2 - x - 1$

to find the intersection of 9/5 and the golden ratio at earth orbit, and get:

9. $x^2 - 10x + 26 = 0$

Equation 9 can be solved with quadratic equation and has the solutions, $(5+i), (5-i)$

These are two complex numbers with real parts 5 and 5, and imaginary parts sqrt(-1) and-sqrt(-1). They are vectors whose sum are:

$(5,1) + (5,-1) = (10,0)$

and whose points are separated by:

$|(5+i)-(5-i)|=2i$

with a modulus of sqrt(5^2+1^2)=sqrt(26)

This generates the triangle with height 5 and base 2i or complex number 5+2i.

The Mandelbrot set is the iteration of a function of complex numbers that generates the fractal given by:

F(z)=z^2+c

If the seed is zero and c is our 5+2i

Then F(0)=(0)^2+5+2i=5+2i

And F(5+2i)=(5+2i)(5+2i)+5+2i=

26+22i

We have generated our 22 cards of the tarot, and our four in that 26-22=4, from 9/5 and the golden ratio.

Furthermore 22=> 2+2=4 and 26=>6-2=4

We have generated four, three times as well, and 3 times four is twelve. It has been Raphael's suggestion that we express all in a simple truth. This might be a start. Also, 22+26=48 and 48/4=12

Four will always represent for me, the fourth planet mars, the one in our solar system that can be terraformed to support human life.

Humans have always liked to quarter things: four directions, north, south, east, and west. Or the Cartesian co-ordinate system in mathematics, used to locate a point with two perpendicularly intersecting axes making four quadrants. It is easy to learn, and effective as it is based on the square because it a regular tessellator. (Tessellate means can tile a surface without leaving gaps, regular that all sides and angles in the shape are congruent.) Likewise nature likes to quarter things. The Earth is rotated to the plane of its orbit by 23.5 degrees. This means on summer solstice, the Sun is directly over-head at noon, the earth receiving the full potency of the Sun's rays, at this latitude. This is La Paz Baja California, Mazatlan Mexico, and Gujarat India. What nature is doing here, is that which Man likes, and uses, and does so because it is practical. 22.5 degrees is the quarter of a way around a half semi-circle, there is only a one-degree discrepancy.

To find when and where the Earth passes through the intersection of 9/5 and golden ratio, which we have shown are (5+i) and (5-i), we use:

a = arctan (y/x) when x>0

a =arctan y/x + pi when x<0

from x+iy = r(cos(a) + isin(a))

r=sqrt(x^2+y^2)

x=5, y=1 yields

arctan 1/5 = arctan 0.2

= 11.3099 degrees

and for (5-i)

we have

-11.3099 degrees

Earth year is 365.25 days

$x/365.25=11.3099/360$

$x360=(365.25)(11.3099)=4130.940975$

$x=11.47483604$ days

Taking winter solstice (December 22) to be our zero point the earth passes through (5+i) on January 2, and (5-i) on December 10.

Note: Since 5+i and 5-i have real and imaginary components, the imaginary component refers to an element at this location in a parallel universe, or tachyon universe as it is called, where mass is negative and travels backwards in time. We as of yet do not know how to penetrate such a plane. Also, keep in mind there is no such thing as absolute space, so 5+i and 5-i are in a frame of reference where the sun is taken as not moving (i.e. the earth is moving with the sun around the galaxy, the galaxy through the universe, and the universe in the multiverse, as so on, the list goes without end. Motion can only be taken as relative (Einstein, special relativity).

Monolith

Arthur C. Clarke's monolith turned out to be a computer put on Earth by extraterrestrials and The Moon to give us an evolutionary nudge when we needed it and to monitor us. It had the dimensions of 1 by 4 by 9, the squares of 1, 2, and 3. 1+4 = 5. The height is 9. I have found that 9/5 (which is 1.8) occurs throughout nature in the areas held most sacred to man down through history, the sun, the moon, gold, silver, water and air:

1. If we compare the mass of air to the mass of water and increase that by a factor of the human body temperature to the freezing temperature of water, we get a value that is 9 compared to 5, which is 1.8.

2. If we compare the mass of an atom of gold to an atom of silver, it is 9 compared to 5 (comparing their molar masses).

3. If we compare the radius of the sun, that is the distance from its center to its surface, to the distance from the center of the earth to the center of the moon, it is 9 compared to five.

9 compared to 5 is nine fifths (9/5) which is equal to 1.8

Glancing at my data tables and find that if we take the distance of the planet Saturn to the sun as 9 (closest approach), then the distance to the planet Jupiter from the sun is five (closest approach). In fact this way of measuring distances puts the earth exactly at 1 unit from the sun. This is interesting, because Jupiter and Saturn, aside from being the "middle children" of the solar system, planets 5 and 6 of a planetary family of 9 or 10 depending on whether or not you consider the asteroid belt a planet that did not form, and anything found beyond Pluto a planetoid, these planets carry the majority of mass of the solar system, significantly, and thus embody most of the dynamics of its formation.

I have also found that the basis of computers and AI (artificial intelligence), which is doped silicon, has the golden ratio in the means of its components. The golden ratio is recurrent throughout life because of the dynamics it has to offer. Doped is silicon, phosphorus, and boron. These are naturally occurring elements, made by nature, namely forged in stars. If P is phosphorus, B is Boron, and Si is silicon (geometric mean by Si):

$$\text{sqrt}(P*B)/Si = \text{sqrt}(30.97*10.81)/(28.09) = 0.65$$

and let us take the harmonic mean between phosphorus and boron and divide it by silicon:

(2*(30.97*10.81)/(30.97+10.81))/28.09 =0.57

Now let us take the arithmetic mean of these two numbers:

(0.65+0.57)/2=0.61

which are the first two digits in the golden ratio.

The golden ratio is 1.618 to three decimal places. Notice that the 2nd and third digits after the decimal are 1 and 8, the two digits in 9/5. The 1 and 6 add up to 7, the average of nine and five, the 6 minus the one is our five, and, the eight plus the one is our nine. So essentially, we have connected the monolith with nature, and computers and artificial intelligence, integrated circuits, transistors, diodes (doped silicon), with the monolith.

ENIGMA

1. The fundamental units are mass, length, and time. When we form ratios between these elements, in the most sacred held aspects of nature, silver, gold, sun, moon, earth, air, water, human body temperature, we get the same ratio, nine fifths.

2. The two most massive planets, saturn and jupiter, are at nine and five units from the sun respectively in their closest approach to the sun, if we take the earth to be one unit from the sun.

3. The most efficient way to divide a circle into equal units (360 degrees) embodies the earth radius and moon radius where the calendar is concerned, which is based on the orbital periods of the moon and earth. The moon and earth could have these periods if the earth and moon were different in size than they are, so this is significant.

4. The nearest stars to our star, the sun, are of brightness's and distances from us that correspond to the sizes of the planets and their number. This is interesting, for example jupiter is the fifth planet and the largest planet in the solar system, the brightest star is the fifth nearest to us, etc …

5. The moon as seen from the earth appears the same size as the sun.

This is the mystical galaxy, pointing out that we are part of an enigma.

Concerning number three (the calendar):

I think this is an incredible thing. By having a seven day week (the Torah) we are able to have twelve four week months where twelve is the most divisible number by whole numbers, for its size. Like this there are twelve new moons between the time it takes for the Earth to go about the sun once, close to it. There are 13, but $(12/13)100=92\%$ accuracy. I did a little number juggling and found that,

Earth-moon distance: $3.84E10 \text{cm} = R$
Earth-sun distance: $1.496E13 \text{cm} = r$
Earth radius: $6.38E8 \text{cm} = E\_r$
Moon radius: $1.738E8 \text{cm} = M\_r$
$(E\_r)/(M\_r) = 11/3$ $r/R = 4675/12$
$((E\_r)(r))/((M\_r)(R)(360)) = 4$

This last equation is significant because, 360 are the degrees in a circle and 4 weeks is a complete revolution of the moon about the earth more or less (time between new moons). 360 degrees are convenient for the units in which to divide a circle, because of its divisible properties. (i.e. it is divisible by 120, 60, 45, 30, and 90, into whole numbers, which are the angles in special triangles.) What this last equation says is that the radius of the earth compared to the radius of the moon, by a factor of the earth-sun distance compared to the earth-moon distance, is equal to 360 times four. It just so happens that after four 360 degree rotations of the earth around the sun, we have a leap year to make the whole calendar work!!!

Also, 360 plus 4 is close to the Earth year. (364/365)100=99.7% accuracy. If you consider that the Earth, Moon and Sun formed from a cloud of gas and dust under its own gravity over perhaps billions of years, this is quite an extraordinary arrangement to have occurred.

THE DATA

The most extraordinary of my findings in my opinion is the occurrence of the proportion of 9 to 5 in nature. I have found that:

1. If we compare the mass of air to the mass of water and increase that by a factor of the human body temperature to the freezing temperature of water, we get a value that is 9 compared to 5, which is 1.8.

2. If we compare the mass of an atom of gold to an atom of silver, it is 9 compared to 5 (comparing their molar masses).

3. If we compare the radius of the sun, that is the distance from its center to its surface, to the distance from the center of the earth to the center of the moon, it is 9 compared to five.

9 compared to 5 is nine fifths (9/5) which is equal to 1.8

I have now glanced at my data tables and find that if we take the distance of the planet Saturn to the sun as 9 (closest approach), then the distance to the planet Jupiter from the sun is five (closest approach). In fact this way of measuring distances puts the earth exactly at 1 unit from the sun. This is interesting, because Juptiter and Saturn, aside from being the "middle children" of the solar system, planets 5 and 6 of a planetary family of 9 or 10 depending on whether or not you consider the asteroid belt a planet that did not form, and anything found beyond Pluto a planetoid, these planets carry the majority of mass of the solar system, significantly, and thus embody most of the dynamics of its formation.

It is further intriguing that air, which I define as the molar weight of the percent of diatomic nitrogen molecules plus the percent of diatomic oxygen molecules at ground level, (nitrogen and oxygen atoms naturally occur as paired atoms) the value is the same considering the percent mass of nitrogen and oxygen in the entire atmosphere because they are the same exact percentages as the percent of diatomic particles at ground level. That is the atmosphere is 21% oxygen and 78% nitrogen by mass and the diatomic oxygen (O2) by number of particles at ground level is 78.7% and (N2) is 21.3% as reported in 1982 in the Handbook of Space Astronomy and Astrophysics by Martin V. Zombeck, Cambridge University Press. These levels of nitrogen and oxygen correspond to levels that are a result of nature's natural regulation process, before we interrupted the cycles of substance regeneration by burning fossil fuels in a way that saturates natures ability to handle the by-products. It would be further interesting, and important to note that the earth naturally is trying to attain a quarter oxygen, three quarters nitrogen exactly in percent of diatomic particles at ground level.

Why is all of this significant? The idea that 9/5 is embodied at the micro level, that is in the masses of atoms of gold and silver, which is related to stellar evolution, the elements were made, forged by the stars, under their gravity, in that furnace whose dynamics are ultimately connected to geometry that was born in the explosion that gave birth to the universe, and is connected to the human body temperature, something regulated by biological processes that took place over billions of years of evolution, and the freezing temperature of water, so crucial to life, and the mass of water and air the chief sustenance's of life. While the same ratio is embodied at the macro level, the proportions of the solar radius and lunar orbital distance which was born by the universe's angular momentum, which is about as complicated as the patterns formed by bubbles in a bath tub swirled by the running water.

All of these factors that form these proportions, so disparately distant in their origins, and so much at the crux of man's awareness down through the ages in poetry, and mysticism, turn out to be related at the most cosmic of levels in a precise empirical way that could only be known today with our telescopes and chemical laboratories, they point to a cosmos that is trying to tell us something significant is happening and that we are a part of something deeply mysterious.

It is worth noting that the natural satellite of Jupiter, Callisto, is 9/5 the density of water.

It is also worth noting that 9/5 is a seventh chord, or transition ratio.

Nine triangles tile to make another triangle with side lengths three times greater (it is a tetractys). 5 points make a trapezoid with three triangles. Thus, 9/5 is an expression of 3 and the earth is the third planet. I have already shown how Jupiter and Saturn, the two most massive planets in the solar system, are in a ratio of 9 to 5 that puts the earth at one unit from the sun. (Aside from showing 9/5 to occur in nature a myriad of ways fundamental to life on earth, in the most revered things, air, water, silver, gold, sun, moon and that thing which is the human common denominator, their normal body temperature). Also, the nine triangles tiled to make a tetractys, divided by the three triangles of five connected points—the trapezoid—is three as well. It is worth mentioning that the average of 9 and 5 is 7. We are talking about the odd numbered rhythms common to arabic (9), flamenco (5), and indian music (7).

calculation of air: $Air=2[(16.00)(0.21)+(14.01)(0.78)]=28.5756$

using molar masses of oxygen and nitrogen, and their percent concentration by mass or percentages of diatomic particles at ground level (the two are the same).

I further find it interesting that Jupiter, the largest planet in the solar system is the fifth planet out of a total of nine planets, another emphasis on 9/5 in the most basic of ways, in nature.

DATA

$$\{(28.5756)/(18.016)\}\{(310)/(273)\}=(197.0/107.9)=1.8$$

and

$$EM/SR=3.84E10cm/6.9599E10cm=0.55$$

EM=earth-moon distance
SR=solar radius

I found a validation for my calculation of Air by Hansen of NASA, the link is in my blog.

2

The Temple

They just unearthed 9 temples on five levels in Italy, put there by a visionary underground who is 57 years old this year. Seven is the average of 9 and five, my discovery of 9/5 in nature and its implications was published in my book "Cosmic Archeology" the year 2007. 7 plus 2 is 9, 7 minus 2 is five. I don't know what is going on here, but I finished this project after my third trip to Italy.

On Hanukkah we Jews burn a candle each day for eight days, but there is a placeholder on the menorah for an extra candle, to light the others with. That is nine holders all together. If the holder for lighting candles is one, then we take the other eight separate, so we have 1.8 which is equal to nine-fifths.

3

The Sacred Volume

Here I demonstrate a relation between the king's chamber, the proportion of sacred ratio 9/5, and the key planets with their formation in the second millennium.

The most extraordinary of my findings in my opinion is the occurrence of the proportion of 9 to 5 in nature. I have found that:

1. If we compare the mass of air to the mass of water and increase that by a factor of the human body temperature to the freezing temperature of water, we get a value that is 9 compared to 5, which is 1.8.

2. If we compare the mass of an atom of gold to an atom of silver, it is 9 compared to 5 (comparing their molar masses).

3. If we compare the radius of the sun, that is the distance from its center to its surface, to the distance from the center of the earth to the center of the moon, it is 9 compared to five.

9 compared to 5 is nine fifths (9/5) which is equal to 1.8

I have now glanced at my data tables and find that if we take the distance of the planet Saturn to the sun as 9 (closest approach), then the distance to the planet Jupiter from the sun is five (closest approach). In fact this way of measuring distances puts the earth exactly at 1 unit from the sun. This is interesting, because Jupiter and Saturn, aside from being the "middle children" of the solar system, planets 5 and 6 of a planetary family of 9 or

10 depending on whether or not you consider the asteroid belt a planet that did not form, and anything found beyond Pluto a planetoid, these planets carry the majority of mass of the solar system, significantly, and thus embody most of the dynamics of its formation.

Nine-fifths (9/5) is embodied by the famous Pythagorean theorem. It is in the first Pythagorean triplet (3,4,5). The Pythagorean theorem, which states the relationship between two sides of a right triangle with its hypotenuse is used to calculate the distance between two points in a plane, or space, given their coordinates in a rectangular coordinate system. It states that one side squared plus the other side squared equals the hypotenuse squared. (3,4,5) is the first triplet that satisfies this relationship. That is 2 squared plus 3 squared equals 5 squared. 3 squared is our 9, 4 squared is 16, and our 5 is squared to make 25. ($3^2+4^2=5^2$). Nine is also the five plus the four. This allows us to associate an angle with nine-fifths, by way of the triangle the triplet describes, and, to find a geometric, or pictorial significance of the ratio.

The ancient Egyptians, associated much significance with this triplet, constructing the King's chamber in the Great Pyramid with these proportions.

Geometric Significance Of nine-fifths, the cosmic equation.

By the Pythagorean theorem

We have $x^2+y^2=h^2$

The first Pythagorean triplet is (3,4,5)

And x=4, y=3, h=5

If

A= area of 3 by 4 rectangle

P= perimeter inscribed in 6 by 8 rectangle

And, n= number of 3 by 4 rectangles in a 6 by 8 rectangle.

Then,

$(A^2)/nP = 9/5$

Is what we have here is a cosmic equation not just for the solar system, but that holds for the atomic world and biological intricacies of life. If we imagine the solar system formed from protoplanetary disc around sun, bodies forming from particles and then herding smaller particles to make distribution of planets as we know them today, I think this equation hold for evolutionary dynamics of solar system and nuclear chemistry dictated by stars to make elements. This picture outlined above will yield much dynamics upon analysis.

A^2=area squared=$P*P*P*P=P^4$

P=perimeter

Number = n

Then since $A^2/P= P^3$= volume

Volume/n=9/5

Saturn orbit (closest approach to the sun) =9 astronomical units
Jupiter orbit (closest approach to the sun)= 5 astronomical units

4th planet = mars

We have:

Volume = (Saturn/Jupiter)Mars

Saturn and Jupiter are the most massive planets in the solar system. Mars is the planet that can be colonized by humans. I have shown that they represent the sacred volume.

April 6 2000 these three planets ushered in the second millennium forming a circle by the moon, 9 degrees across.

4

Calculation of Sacred Volume

E_r=1= Earth radius

M_r=0.532= Mars radius

J_r= 11.27= Jupiter radius

S_r=9.44= Saturn radius

$V = 4/3 (pi)r^3$

$= 4/3(3.1)(11.27)^3 = 5996$

$=4/3(3.1)(9.44)^3 = 3524$

$=4/3((3.1)(0.532)^3 = 0.6307$

Volume = (Saturn/Jupiter)mars

$= 3524/5996(0.6307) = 0.37$ cubic earth radii

Approximately the same number as mercury radius and mercury distance from sun in earth units.

5

Sacred Volume Analyzed, Sacred Length Obtained, Dynamic Structure Of Solar System Revealed.

I calculated the sacred volume to be 0.37 cubic earth radii, which I have noticed coincides with mercury distance from the sun, and mercury radius in earth units. Since volume is in cubic unit lengths, and radius and distances in earth radius and earth-sun distances are lengths, we take the cube root of our sacred volume to get the length of one side of a cube with the sacred volume. It is 0.72 earth radii. We will call it the sacred length, it is the venus-sun distance in earth units, where venus is the planet after mercury, or the second planet. This divides into the mercury orbit 0.51 times. We invert that to compare this cube root to to mercury orbit, and get 1.96. This number is close to 1.8, what I call the cosmic number, or sacred number, nine fifths, which may be worth noting, considering the grand scale of the solar system and the many factors that influenced its formation. Mercury is the closest planet to the sun, the first planet, and is very hot, it is named for a Greek god, as are all the planets, and bears the same name as the element used in thermometers to measure temperature. Mercury was the fleet footed messenger of the gods.

6

Recent Developments and Summary

I have found nine-fifths (9/5), which is equal to 1.8 in the most sacred aspects to man, gold, silver, moon, sun, water, air, human body temperature, and have have derived from this sacred ratio, or cosmic number, the cosmic volume by expressing it geometrically. That sacred volume is:

(Saturn/Jupiter)mars=0.37 cubic earth radii

Where Saturn and Jupiter are the most massive planets in the solar system, and Mars is the planet that can be colonized by humans. Taking the cube root of 0.37 gives the sacred length, it is equal to the distance of the planet Venus from the sun in earth units (astronomical units), which is the second planet. The 0.37 is the mercury-sun distance in earth units, and the mercury radius in earth units. Mercury is the first planet.

Saturn is at 9 earth units in its closest approach to the sun, Jupiter at five.

Thus we see a divine structure for the first six planets, Mercury, Venus, Earth, Mars, Jupiter, Saturn. This leaves Uranus and Neptune, planets seven and eight respectively. For now we won't consider Pluto a planet, as its orbit is well inclined to the plane of the solar system.

7

The Cosmic Fulcrum

I have found that 9/5 occurs in many mysterious ways in nature, including Saturn's distance from the sun of 9 (closest approach) putting Jupiter's at five (closest approach), and the earth at one. Let us consider a balance. If two weights of equal mass are placed on opposite ends of a balance, then to be balanced the fulcrum must be placed at the center. But if one mass is increased, then the fulcrum will have to be moved closer to the heavier end if the balance is not to tip. This is the Law of Levers. It states that the masses are inversely proportion to their distances from the fulcrum. To write it mathematically,

M1/M2=L2/L1

Or, equivalently

(M1)(L1)=(M2)(L2)

If we consider the time when Saturn and Jupiter, the two most massive bodies in the solar system are in opposition, that is when they are on opposite sides of the sun and facing one another, then by the above information they are separated by 14 units. Jupiter is 3.34 times as massive as Saturn. By our law of levers above, that means

L1/L2=3.34

L1+L2=14

Thus

4.34L2=14 and

L2=3.2258

14-3.22=10.77=L1

Mars orbital distance is 1.5

The saturn-fulcrum distance minus the mars orbital distance is: 10.77-1.5=9.2~9 is saturn's closest approach to the sun.

Mars, then, is the cosmic fulcrum. It is the one planet worth terraforming for colonization. It is then aligned metaphysically with its physical characteristics.

8

Nine-Fifths

I have been trying to connect various data about the solar system and atomic world in hopes that a picture will start to form regarding the paradox of existence. A whole lot of stuff has cropped up and this one in particular demonstrating the relationship between the sun, moon, air, water, gold, silver, the human body temperature, and the freezing temperature of water, is intriguing to me because not only are silver and gold the most precious metals of ceremonial jewelry, the moon and sun are prominent in poetry down through the ages and water and air are the sustenance of life. If one considers the way a solar system might form, this is extraordinary.

(m_a/m_w)(human body temp/freezing of water)

$=(Au/Ag)$

$=SR/EM$

$=1.8$

In words: the mass of air by the mass of water times the human body temperature by the freezing of water is the same as gold over silver is the same as the solar radius over the earth-moon distance.

Since the atmosphere is more or less 21% oxygen gas (O_2) and 78% nitrogen gas (N_2)

$Air=2[(16.00)(0.21)+(14.01)(0.78)]=28.5756$

This is at sea level. 1.8 is the magic number. Now that is equal to 9/5. The five sided pentagon contains the golden ratio. What can we say about the number 9? All reciprocal intervals add up to 9. I calculate 9/5 to be a musical interval of a seventh. Nature is playing a seventh chord in a very interesting way. The seventh chord is a lead in chord for a change in major to minor 4 steps higher. But 9/5 is perfectly a seventh in a non-tempered system such as that used in Indian music from India, or perhaps even in Middle Eastern music.

I found another instance of 1.8 (9/5). The moon of Jupiter called Callisto is 1.8 times as dense as water. Some facts about it are that it is Jupiter's second largest moon and is the third largest in the solar system.

DATA

$\{(28.5756)/(18.016)\}\{(310)/(273)\} = (197.0/107.9) = 1.8$

and

$EM/SR = 3.84E10cm/6.9599E10cm = 0.55$

1.8 is straddled by the golden ratio and 2 by point 2 units on either side it, if we round the golden ratio to 1.6, the value being 1.618 to three decimal places. While the golden ratio divides a line such that the whole to larger is the same as the larger to the lesser, 2 is unity on either side. Notice that to three decimal points the residual of rounding the golden ratio to one decimal place is 18, the two numbers of our one point eight.

If 0.21 and 0.78 represent the mass percentages of oxygen and nitrogen respectively in the atmosphere in the above calculation, this is what I call "the essence of air." Interesting that these two numbers are nearly the percent of O2 and N2 particles respectively at ground level. I calculated from the Handbook of Space Astronomy and Astrophysics by Martin V.

Zombeck, Cambridge University Press 1982, that considering just N2 and O2 we have precisely 78.7% N2 and 21.3% O2 of particles at ground level. It gives the smallest difference in our cosmic number. Both answers are 1.8 rounded to one decimal place.

9

outline of factors used in nature study

Humans settled in communities to farm and ranch, after hunting and gathering, perhaps 12,000 years ago, or more. Since that time, and long before it, the sun has been burning at the same temperature, more or less, the climate of the Earth varying in the Northern and Southern Hemispheres due mostly to their inclinations towards and away from the sun which alternate as the Earth goes around the sun, yearly (the earth is inclined to its orbit by about 23.5 degrees.) There are other factors that effect climate on earth, like ocean currents. The distances of the planets from the sun have been constant over geologic time, or more, as they orbit the sun. The elements are the same from sample to sample on earth, for all practical purposes. That is a mole of magnesium, for instance, weighs the same whether found in Alaska or Africa. The masses of the planets are constant. We mostly compare the molar masses of the elements (a mole being 6.02E23 atoms), and find where these ratios compare to the ratios formed between

1. Solar luminosity over a year and planetary kinetic energy (energy of motion around the sun).
2. Planetary masses
3. Separations between planets and the sun

We consider history, as it has been forged by the elements like copper for tools and much later for electrical wire. We are considering such relationships while the Titius-Bode Rule holds, since longer than the advent of man, which is an equation that predicts the distribution of the planets about the sun. It is geometric in nature.

10

On the Ingredients of Life

An interesting family of substances is methane (CH_4), ammonia (NH_3) and water vapor (H_2O). Methane is tetrahedral in structure, a carbon atom surrounded by 4 hydrogens. Ammonia is trigonal pyramidal, a nitrogen atom surrounded by 3 hydrogen atoms, and water vapor is triangular, or bent, an oxygen atom surrounded by two hydrogens. These represent stable structural systems as they are all systems of triangles, which are the only stable polygons. These substances combined under energy with hydrogen gas form amino acids, the building blocks of life. The core atoms of these molecules, carbon, nitrogen, and oxygen, are all in period two of the periodic table and follow directly one after the other, and are all in amino acids, the hydrogen as well. It is a hypothesis of astrobiology that amino acids formed in the protoplanetary cloud before the earth ever formed. In this sense we may have our origins in deep space. Is what I mean by structural systems is that there are only three structural systems; the tetrahedron, the octahedron, and the icosohedron. They are the only stable solids, that is non-collapsing flex corners whose faces are triangles. Most compounds are something other than these, like pentagons with linear off shoots for example, that comprise the wrong number of atoms to make a "solid" unit, and I mean solid as in the pythagorean solids, the geometric term. Both methane and ammonia make different variations on the tetrahedron, a pythagorean solid. When plants perform photosynthesis, they combine carbon dioxide with water and release oxygen. The reaction is:

$$CO_2 + 2H_2O \longrightarrow CH_2O + O_2 + H_2O$$

As can be seen a sugar is made. Important to most plants to do this is Nitrogen. Nitrogen (N_2) is the most abundant gas in the earth atmosphere, comprising about 78.03% of it. We now calculate the molecular masses of these special gases:

$CH_4=(12.01+4(1.01))=16.05$
$NH_3=(14.01+3(1.01))=17.04$
$CO_2=(12.01+2(16.00))=44.01$
$H_2O=(2(1.01)+16.00)=18.02$
$N_2=(14.01+14.01)=28.02$
$O_2=(16.00+16.00)=32.00$

We now form some ratios between these molecular masses:

$(O_2)/(CH_4)=32.00/16.05=1.992\sim2$
$(NH_3)/(CH_4)=17.04/16.06=1.061\sim1$
$(CO_2)/(O_2)=44.01/32.00\sim1.4=sqrt(2)$
$(CO_2)/(N_2)=44.01/28.02\sim1.6=(sqrt(5)+1)/2=phi$
$(O_2)/(H_2O)=32.00/18.02=1.776\sim sqrt(3)$

Notice that these values are given by the sequence:

$|2cos(pi/n)|$ n=(1,2,3,4,5,6)(pi/n)radians
Observe:
$2=|2cos(pi)|$
$0=|2cos(pi/2)|$
$1=|2cos(pi/3)|$
$sqrt(2)=|2cos(pi/4)|$
$(sqrt(5)+1)/2=phi=|2cos(pi/5)|$
$sqrt(3)=|2cos(pi/6)|$

Geometrically sqrt(2) is the ratio of the side of a square to its radius. Phi is the ratio of the chord of a regular pentagon to its side. Sqrt(3) is the ratio of the side of an equilateral triangle to its radius, and 1 is the ratio of the side of a regular hexagon to its radius. The square, the regular hexagon and the equilateral triangle are the tessellating regular polygons.

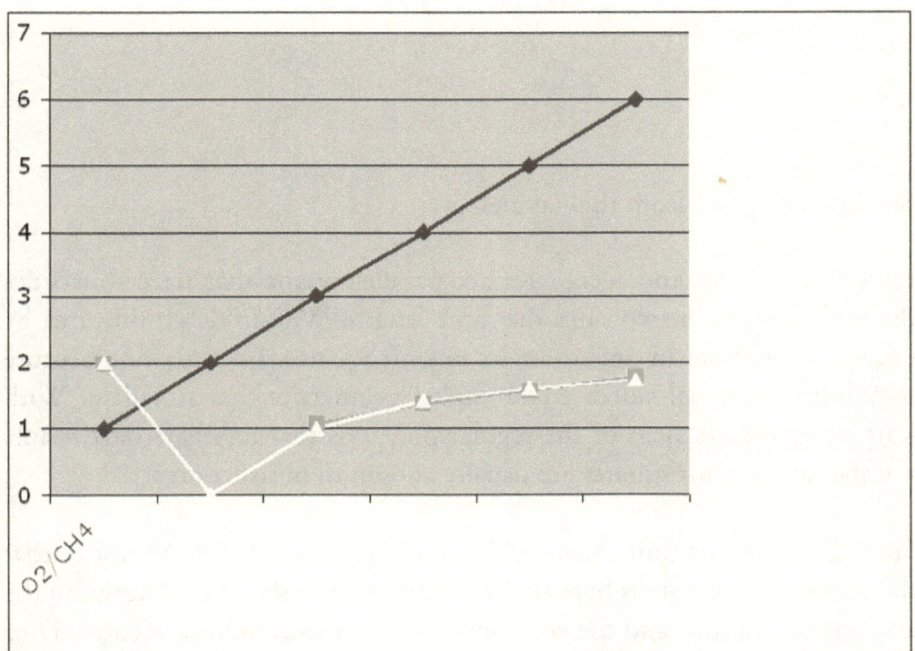

11

Math System

Formulas Derived from the Parallelogram

Remarks. Squares and rectangles are parallelograms that have four sides the same length, or two sides the same length. We can determine area by measuring it either in unit triangles or unit squares. Both are fine because they both are equal sided, equal angled geometries that tessellate. With unit triangles, the areas of the regular polygons that tessellate have whole number areas. Unit squares are usually chosen to measure area.

Having chosen the unit square with which to measure area, we notice that the area of a rectangle is base times height because the rows determine the amount of columns and the columns determine the amount of rows. Thus for a rectangle we have:

$A=bh$

Drawing in the diagonal of a rectangle we create two right triangles, that by symmetry are congruent. Each right triangle therefore occupies half the area, and from the above formula we conclude that the area of a right triangle is one half base times height:

$A=(1/2)bh$

By drawing in the altitude of a triangle, we make two right triangles and applying the above formula we find that it holds for all triangles in general.

We draw a regular hexagon, or any regular polygon, and draw in all of its radii, thus breaking it up into congruent triangles. We draw in the apothem of each triangle, and using our formula for the area of triangles we find that its area is one half apothem times perimeter, where the perimeter is the sum of its sides:

A=(1/2)ap

A circle is a regular polygon with an infinite amount of infitesimal sides. If the sides of a regular polygon are increased indefinitely, the apothem becomes the radius of a circle, and the perimeter becomes the circumference of a circle. Replace a with r, the radius, and p with c, the circumference, and we have the formula for the area of a circle:

A=(1/2)rc

We define the ratio of the circumference of a circle to its diameter as pi. That is pi=c/D. Since the diameter is twice the radius, pi=c/2r. Therefore c=2(pi)r and the equation for the area of a circle becomes:

A=(pi)r^2

(More derived from the parallelogram)

Divide rectangles into four quadrants, and show that

A. (x+a)(x+b)=(x^2)+(a+b)x+ab

B. (x+a)(x+a)=(x^2)+2ax+(a^2)

A. Gives us a way to factor quadratic expressions.

B. Gives us a way to solve quadratic equations. (Notice that the last term is the square of one half the middle coefficient.)

Remember that a square is a special case of a rectangle.

There are four interesting squares to complete.

1) The area of a rectangle is 100. The length is equal to 5 more than the width multiplied by 3. Calculate the width and the length.

2) Solve the general expression for a quadratic equation, $a(x^2)+bx+c=0$

3) Find the golden ratio, a/b, such that $a/b=b/c$ and $a=b+c$.

4) The position of a particle is given by $x=vt+(1/2)at^2$. Find t.

Show that for a right triangle $(a^2)=(b^2)+(c^2)$ where a is the hypotenuse, b and c are legs. It can be done by inscribing a square in a square such that four right triangles are made.

Use the Pythagorean theorem to show that the equation of a circle centered at the origin is given by $r^2=x^2+y^2$ where r is the radius of the circle and x and y the orthogonal coordinates.

Derive the equation of a straight line: $y=mx+b$ by defining the slope of the line as the change in vertical distance per change in horizontal distance.

Triangles

All polygons can be broken up into triangles. Because of that we can use triangles to determine the area of any polygon.

Theorems Branch 1

1. If in a triangle a line is drawn parallel to the base, then the lines on both sides of the line are proportional.

2. From (1) we can prove that: If two triangles are mutually equiangular, they are similar.

3. From (2) we can prove that: If in a right triangle a perpendicular is drawn from the base to the right angle, then the two triangles on either side of the perpendicular, are similar to one another and to the whole.

4. From (3) we can prove the Pythagorean theorem.

Theorems Branch 2

1. Draw two intersecting lines and show that opposite angles are equal.

2. Draw two parallel lines with one intersecting both. Use the fact that opposite angles are equal to show that alternate interior angles are equal.

3. Inscribe a triangle in two parallel lines such that its base is part of one of the lines and the apex meets with the other. Use the fact that alternate interior angles are equal to show that the sum of the angles in a triangle are two right angles, or 180 degrees.

Theorems Branch 3

1. Any triangle can be solved given two sides and the included angle.

c^2=a^2+b^2-2abcos(C)

2. Given two angles and a side of a triangle, the other two sides can be found.

a/sin(A)=b/sin(B)=c/sin(C)

3.Given two sides and the included angle of a triangle you can find its area, K.

K=(1/2)bc(sin(A))

4.Given three sides of a triangle, the area can be found by using the formulas in (1) and (3).

Question: what do parallelograms and triangles have in common?
Answer: They can both be used to add vectors.

Trigonometry

When a line bisects another so as to form two equal angles on either side, the angles are called right angles. It is customary to divide a circle into 360 equal units called degrees, so that a right angle, one fourth of the way around a circle, is 90 degrees.

The angle in radians is the intercepted arc of the circle, divided by its radius, from which we see that in the unit circle 360 degrees is 2(pi)radians, and we can relate degrees to radians as follows: Degrees/180 degrees=Radians/pi radians

An angle is merely the measure of separation between two lines that meet at a point.

The trigonometric functions are defined as follows:

cos x=side adjacent/hypotenuse

sin x=side opposite/hypotenuse

tan x=side opposite/side adjacent

csc x=1/sin x

sec x=1/cos x

cot x=1/tan x

We consider the square and the triangle, and find with them we can determine the trigonometric function of some important angles.

Square (draw in the diagonal): cos 45 degrees =1/sqrt(2)=sqrt(2)/2

Equilateral triangle (draw in the altitude): cos 30 degrees=sqrt(3)/2; cos 60 degrees=1/2

Using the above formula for converting degrees to radians and vice versa:

30 degrees=(pi)/6 radians; 60 degrees=(pi)/3 radians.

The regular hexagon and pi

Tessellating equilateral triangles we find we can make a regular hexagon, which also tessellates. Making a regular hexagon like this we find two sides of an equilateral triangle make radii of the regular hexagon, and the remaining side of the equilateral triangle makes a side of the regular hexagon. All of the sides of an equilateral triangle being the same, we can conclude that the regular hexagon has its sides equal in length to its radii.

If we inscribe a regular hexagon in a circle, we notice its perimeter is nearly the same as that of the circle, and its radius is the same as that of the circle. If we consider a unit regular hexagon, that is, one with side lengths of one, then its perimeter is six, and its radius is one. Its diameter is therefore two, and six divided by two is three. This is close to the value of pi, clearly, by looking at a regular hexagon inscribed in a circle.

The sum of the angles in a polygon

Draw a polygon. It need not be regular and can have any number of sides. Draw in the radii. The sum of the angles at the center is four right angles, or 360 degrees. The sum of the angles of all the triangles formed by the sides of the polygon and the radii taken together are the number of sides, n, of the polygon times two right angles, or 180 degrees. The sum of the angles of the polygon are that of the triangles minus the angles at its center, or A, the sum of the angles of the polygon equals n(180 degrees)-360 degrees, or

A=180 degrees(n-2)

With a rectangular coordinate system you need only two numbers to specify a point, but with a triangular coordinate system—three axes separated by 120 degrees—you need three. However, a triangular coordinates system makes use of only 3 directions, whereas a rectangular one makes use of 4.

A rectangular coordinate system is optimal in that it can specify a point in the plane with the fewest numbers, and a triangular coordinate system is optimal in that it can specify a point in the plane with the fewest directions for its axes. The rectangular coordinate system is determined by a square, and the triangular coordinate system by an equilateral triangle. They are the basis for many mosaics in Moorish castles, such as those in the Alhambra in Spain.

From the Physics Notebook of Ian Beardsley
F=ma M=mv v=x/t
F=Force M=momentum m=mass v=velocity x=distance t=time a=acceleration
M=mv=m(x/t) a=dv/dt a(dt)=dv v=at v=dx/dt dx/dt=at dx=at(dt) x=(1/2)at^2
x=x_0+vt+(1/2)at^2
int[x^n] 0 to x =(x^(n+1))/(n+1) and (d/dx)x^n=nx^(n-1)

K=kinetic energy U=potential energy C=constant

K=(1/2)mv^2 U=mgy h=height

K+U=C mgh=mgy+(1/2)mv^2 or (1/2)m(v_0)^2=U+K where v_0=initial velocity

Work=W=Fx and U=-W

Thus work is the distance traveled or moved by the component of the force in that direction, and potential energy is the negative of the work. Use the definition for work and the chain rule for derivatives to show that kinetic energy (energy of motion) is as given above.

The chain rule is:

dv/dt=(dv/dx)(dx/dt)

A ball rolling on an incline will stay in motion until it attains the same height on another incline facing the first, even if the inclinations of the two inclines are not the same. If there is no second incline, the ball will never attain the original height and will therefore continue to roll forever, unless otherwise acted on by a force, like friction. For every force there is an equal but opposite reaction.

Notice that:

mgh=(1/2)m(v_0)^2

12

Derivation of the Golden Ratio

The golden ratio is the ratio such that the whole portion to the larger portion equals the larger portion to the lesser, or in other words, we have that a/b=b/c when a=b+c.

$(a/b) = (b/c)$ when $a = b + c$

$(a/b) = (b/c)$

$ac = b^2$

$a = b + c$

$c = a\text{-}b$

$a(a\text{-}b) = b^2$

$a^2\text{-}ab = b^2$

$a^2\text{-}ab\text{-}b^2 = 0$

$(a^2/b^2)\text{-}(a/b)\text{-}1 = 0$

The last equation is a quadratic in a/b.

$(a^2/b^2)\text{-}(a/b) = 1$

$(a^2/b^2)-(a/b) + (1/4) = (5/4)$

$((a/b)-(1/2))^2 = (5/4)$

$(a/b)-(1/2) = (sqrt(5))/2$

$a/b = (sqrt(5) + 1)/2$

13

Experimental Science

date	time	moon set	moon rise	set ratio	incline set	incline rise
9/24/04	1:20AM	233 deg W		0.647	20 degrees	
9/26/04	3:10AM	231 deg W		0.641	22 degrees	
9/27/04	4:45AM	250 deg W		0.694	4 degrees	
9/27/04	7:35PM		111 deg SE			9 degrees
9/28/04	8:10PM		96 deg SE			14 degrees
9/28/04	8:17PM		93 deg SE			20 degrees
9/29/04	8:40PM		87 deg NE			17 degrees
9/30/04	8:58PM		82 deg NE			9 degrees

"Incline set" is degrees of moon above horizon when I measured it setting.

"Incline rise" is degrees of moon above horizon when I measured it rising.

The "set ratio"is the ratio of degrees west the moon set compared to 360 degrees. I hypothesize the moon will set at the golden angle when it sets its furthest south (222 degrees west, which is a "set ratio"of 0.618, the golden ratio.)

Position as measured with a compass is different than position as measured with a telescope. A telescope measures "right ascension" which are projections of longitude lines onto the celestial sphere, whereas a compass measures along the horizon in an apparent plane clockwise starting at north. I have also made a sextant for measuring inclination. This is different as well from the astronomer's declination, which is measured by projections of latitude lines onto the celestial sphere. Thus these are not ephemeral positions.

Now it is entirely possible that ancient cultures might have noticed times when the moon set at an angle of rotation clockwise around the horizon from north that is the same angle of rotation around the stem of a plant by consecutive leaves, and identified this as approximately 2/3, an approximation to the golden angle. We shall see just when the moon sets at the golden angle this year, if it has not already. And it is looking like it will do it this year.

The moon should measure the same setting and rising positions from any different positions on the Earth, within reason, the earth is curved, at the same time because the angle made with any two positions on the Earth, and the moon, will be very small, because the moon is so far away compared to any change in position on the Earth. Therefore these measurements are position independent, within the accuracy of my compass and sextant, which are both plus or minus one degree. See the picture below.

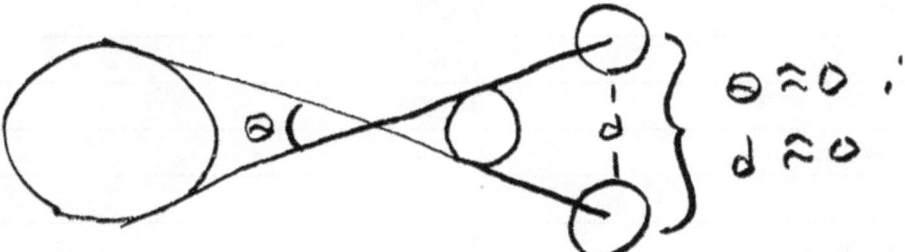

I used an accurate military compass with viewing slit and cross hair to make the measurements. There is a built in magnifying glass to read position. The compass position parameters above are normalized to account for the difference between true north and magnetic north, which varies with latitude and longitude. At my latitude of 34 degrees north, 117 degrees west, the magnetic declination, or degrees east that magnetic north is of true north is 13 degrees. Therefore 13 degrees were subtracted from all of my compass readings. I made my sextant from a protractor, drilling a hole in the middle of the straight edge part from which to hang a pointer that remains pointing at the Earth's center by gravity as the readings rotate through it.

It is a curious thing about the moon that, as seen from the Earth, it appears to be the same size as the sun, nearly enough that when it passes between ourselves and the sun there is a near perfect eclipse, which allows us to observe the outer thin atmosphere of the sun, that is otherwise impossible to observe, because of the brightness of the sun's main body. When we take into account that the most spectacular meteor shower of the year, the perseids, is heralded by the heliacal rising of the brightest star in the sky, Sirius, here in Southern California on August 12, then things become quite interesting. Here in the mountains that surround this valley, the San Gabriel Mountains, occurs a rare deep blue gemstone, Lapis Azuli, which is protected by law. The only other place in the world where it occurs, are the mountains of Afghanistan.

Moon	
mass	7.35E25g
radius	1,738km
density	3.34g/mL
gravity	0.166g
day	27.322d
inclination	6.68 deg

Around this house alone over the past couple of months I have been approached by three immature red tail hawks that landed in the pine tree next to me, by one black rabbit, and by a family of raccoons several evenings. I have seen a coyote in the foothills, and there are plenty of squirrels all over town. Western scrub jays and crows are highly abundant. In the mountains I have seen herds of big horn sheep.

Native to this valley, and throughout the state, is the golden poppy wildflower, which does well in dry, harsh climates. It even flourishes in great abundance in the Mojave Desert on the other side of the mountains. There is a golden poppy reserve there, in Antelope Valley.

On 10/22/04 I measured the moon to set at the golden ratio, which happened to coincide with the village venture in Claremont, California, a sort of celebration of the fall equinox.

14

The Structure of Matter

The numbers for atoms, or chemical elements and compounds, are relative masses, which vary because the different kinds of matter are made up of units that have varying numbers of identical particles. These identical particles are three: protons, neutrons, and electrons. Protons and neutrons are almost the same mass, electrons are lighter. Neutrons have no charge, but protons are positively charged and electrons are negatively charged. Is what we mean by charged is that the particle has associated with it an electric field, which means that it can attract or repel particles that are charged. We say that opposite charges attract, and like charges repel. The number of particles that combine to make an atom, or element, determines its characteristics and quality. Elements are atoms, that cannot be reduced further unless through a so-called nuclear reaction. Compounds are combinations of atoms, and can be reduced to atoms through a chemical reaction. Nuclear reactions require allot of energy, as in what is occurring at the center of the sun, and stars. Chemical reactions can occur here on earth, in the laboratory, by mere mixture of the right elements, or by applying a flame. Most matter is neutral because it has an equal number of protons and electrons. If it has varying numbers of neutrons, we say that an element with more is simply a heavier version of the same. Elements that vary in weight alone are called isotopes. If an atom, or element, has an unequal number of protons and electrons, then it has a net charge and we call it an ion. More electrons and it is a negative ion, and less it is a positive ion. Electricity is the flow of electrons through a wire that is often an element such as copper or aluminum. Silver is the most conductive metal. Aside from being a metal, elements can be semimetals, or non-metals. Matter can exist in four states, solid, gas, liquid, or plasma. Some elements at earth temperatures are gases, like those that comprise the atmosphere. Though an element like Iron is a solid at earth temperatures, we can heat it in a forge and it becomes a liquid. It becomes a liquid when heated because adding energy makes its atoms further apart, and thus more movable. When we speak of particles, we are speaking of sub-atomic particles. Sub-atomic particles are actually composed of even smaller particles called quarks. We can use a series of magnets to accelerate sub-atomic particles to velocities that are a significant percentage of light velocity, and smash them into targets, to study how matter interacts. Photons are massless particles of light. Light can behave like an electromagnetic wave, or like a particle, depending on what we do with it. This is called wave-particle duality.

The elements themselves were made by stars the heavier elements having been made by combining with the lighter elements under the pressure of the star by its own gravity. Energy is released in this nuclear reaction, a process called "fusion", and it is the outward force from this that causes a balance for the star with the inward force.

15

Stars Connected to Planets

Let the closest star, Rigel Kentaurus, also called alpha centauri, be a metaphor for the earth, in that it is the closest star to the sun, and the third brightest in the sky and the earth is the only planet brimming with life and is the third planet from the sun. Thus we now consider the largest planet in the solar system (Jupiter) and this takes us to the brightest star in the sky, Sirius, alpha canes major, it is the fifth nearest star and Jupiter is the fifth planet. But let us associate with Jupiter as well Vega, it is the fifth brightest star. Sirius is 8.7 light years distant, and Rigel Kentaurus is 4.34 light years distant. 8.7/4.34=2.00 ... If "the surface of the earth is the shore of the cosmic ocean" as Carl Sagan said, then the constellation Bootes carries its importance in the fact that it is "the boatman". The brightest star in that constellation is Arcturus, which happens to be the fourth brightest star in our galaxy. The fourth planet is mars, and as it so happens, it is the only planet in our solar system we can colonize, in that mercury is so close to the sun that it is far too hot, venus has the same problem, but mars is the next planet after the earth, and the rest being gas giants, you can't really set foot on them. That is, we say that mercury, venus, the earth and mars are the terrestrial planets.

Pluto is a terrestrial planet as well, but cold because of its distance to the sun, though they are considering calling it a planetoid because of its size.

16

Medicine Wheels

Making a "medicine wheel" is easy. As the earth orbits the sun, the sun appears to swing from south to north and back south again over the course of a year. The southern most position marks the beginning of winter (winter solstice). The northern most position marks the beginning of summer (summer solstice). As the sun travels north, we come out of winter, when it is halfway there, that is the beginning of spring (spring equinox). After the sun reaches its northern most position, it turns around and heads south again. When it reaches the halfway point on its journey back south, that is the beginning of fall (fall equinox). The four seasons, then, can be marked by these three positions of the sun. To make a medicine wheel, an old way of predicting the seasons, align two stones to point at these key positions of the sun, when it sets. The best way is not with compass and line, but simply to place the stones over the course of a year, when the sun is in each of these positions.

17

Astronomy Timeline

A.D. 100's Ptolemy offers the idea of epicycles to explain the retrograde motion of the planets.

About 310B.C.-230B.C. Aristarchus determines the distance to the moon and that the sun is very far away using parallax.

(276–195?B.C) Eratosthenes determines the circumference of the earth by the angle cast by the shadow of a stick and the absence of a shadow in a well in another location.

1473–1543 Copernicus puts forward his model of a solar system where the earth is not at the center of the universe, but that goes around the sun along with the other planets. The model explains the retrograde motion of the planets.

1564–1642 Galileo discovers four natural satellites orbiting Jupiter, and thus verifies the idea of Copernicus that the Earth is not at the center of the universe.

1571–1630 Based on the observations of Tycho Brahe, Kepler formulates the laws of planetary motion.

1642–1727 Newton makes his universal law of gravitation from which Kepler's laws can be derived. He invents differential calculus and integral calculus, the latter simultaneously with Liebnitz.

18

Observing the Heavens

If a shooting star (meteor) is big enough, you get what is called a "fireball". While working at Pine Mountain Observatory, eyes skyward all night, every night during the summer, and on winter weekends, I have seen several.

As a comet orbits the sun, it leaves behind debris in its orbit, small rocks and pebbles. When the earth flies through the orbit annually, it flies into the rocks and pebbles, as they burn up in the atmosphere, the friction excites the atoms of the air, ionizing it, so as to cause it to emit light, leaving a luminous trail, commonly called a shooting star. Shooting stars can happen anytime, but when we fly through the orbit of a comet, we have a meteor shower. If the meteor is big enough as not to burn up before hitting the earth, then we can often find the rock, called a meteorite. In August, we fly through the orbit of a particular comet, that provides for what is usually the most spectacular meteor shower, the perseids. It is called the perseids meteor shower, because the meteors tend to emanate from the constellation perseus, because that is the direction the earth is moving into, during that time of the year. This point is called "the radiant", and if you point a camera at it and leave the shutter open for an hour or so, you will get an image of many trails coming from a point.

When I was at Pine Mountain, amateur astronomers would set up their telescopes, and shows tourists celestial bodies with their often homemade telescopes. We showed tourists celestial bodies through the observatory telescopes as well. There are amateur astronomy clubs throughout the world, that aside from holding star parties, hold workshops on making

51

your own telescopes. Amateur Astronomy is one of the few fields where amateurs make contributions to a professional field, discovering comets, and providing a greater observing power for an otherwise immense universe, for which not nearly enough observatories exist to monitor everything.

19

Triangulation

I have sought to connect symbols too, that derive from data that is available, but well hidden and extraordinarily esoteric as it derives from such places as my old stomping grounds in the least considered of possible places by scholars and mystery detectives, the remote boonies, in hope to score on some real exciting gems. Here is one:

We apply the triangulation of Chris Darrow, where three relevant locations are chosen on a map, are connected and the associated power spots pertaining to the corners are revealed. Bear in mind that a triangle is the minimal structure that encloses an area.

I have chosen Nehalem, Oregon as it is where I estimate that standard temperature and pressure occurs most frequently, that is freezing temperature of water and 1 atmosphere of pressure. I have chosen as a second point Red Bluff, California as its annual average temperature is the mean of freezing and the normal body temperature, 37 deg C. I have chosen as the third point Bend, Oregon, as it is "land's end" in the high desert the last city before the vast expanse to Idaho, nearly. Connecting these points we have a triangle who's "center of gravity" seems to be the Cascade Mountain range Summit at Crescent Lake. Red Bluff is in the agricultural hub of California, I think is the biggest producer of well, produce in the world.

20

Human Destiny

It is quite possible that events on Earth are in sync with the the yearly passage of the Earth around the sun and, cosmic constants like the speed of light. Mankind is Myth, observe:

Iron was not separated from its ore until about 3500 B.C. in the Middle East, but around 4,000 B.C. iron was obtained from meteorites, rocks that came to earth from space, and used to make spear points and ornaments, in Sumeria and Egypt. That would have been 6,000 years ago from today, more or less, 2005, this the beginning of the Space Age (the first private venture ship leaves Earth atmosphere on June 21 2004, called Space-ShipOne). We now calculate how many miles the earth has traveled over that amount of time in its roughly 365-day journey around the sun.

The Earth orbit being nearly circular, its eccentricity varying over time from 0.0 to 0.06, it is sufficient to use c=2(pi)r to calculate its circumference. The average distance of the Earth from the sun is 1.495979E13cm.

(1.495979E13cm)(m/100cm)(km/1000m)=1.495979E8km

(1.495979E8km)(3.14)(2)=9.39E8km

(9.39E8km/year)(6000years)=5.634E12km

One light year (ly) is 9.460530E17cm, it is the distance light has traveled in one revolution of the Earth about the sun.

(9.46E17cm)(m/100cm)(km/1000m)=9.46E12km

5.634/9.46=0.59~0.60=3/5

Thus the distance the earth has traveled around the sun in the last 6,000 years, since iron was first crafted into spear points and so forth from meteorites, compared to the current distance light travels in one year is nearly 3 to 5. Three and five are two consecutive Fibonacci numbers. The ratios between adjacent terms in the Fibonacci sequence are approximations to the golden number, or phi as it is called, which come closer and closer to the value as the numbers increase in value. Three and five are the fifth and sixth terms respectively.

I am no historian, and I have no idea of how accurate the date I have mentioned for these iron spear points and ornaments can be, but let us reverse the concept here and say that iron spear points were fashioned from meteorites 6,000 years ago because in that time from the beginning of the Space Age the Earth has traveled a fraction of a light year equal to the golden ratio. In other words there is interconnectivity between all things, between events on earth that are human, even animal or plant in general, and what is going on through out the cosmos. Whether or not the first spear points were made from meteorites 6,000 years ago, it can be said that it was the beginning of the Egyptian calendar (4241 BCE the Egyptian calendar was established). What exactly was going on after the Earth had traveled around the sun exactly 0.618, the golden number, of a light year after this date? Here it is calculated.

(0.618)(9.46E12km)=5.8E12km

(9.39E8km/year)(x)=5.8E12km

x=6,177 years

6,177 years-4,241 years=1936 A.D.

Around this time Alan Turing published his paper that founded the field of artificial intelligence (1937), and Theodosius Dobzhansky explained how evolution works (1937).

Man was made by evolution, artificial intelligence would be man making the mirror image of himself.

21

Worth Noting

There are about 100 moons to the earth from the moon, and about 100 suns to the earth from the sun. This allows for a perfect solar eclipse. If the solar radius is 9, the earth-moon distance is five. 100 times nine-fifths are 180, the number of degrees in the semi-circle eclipsed face. 180 degrees makes a full circle of 360 degrees. 360 degrees are convenient for dividing the circle because it allows for whole number division by angles in special triangles, 30, 45, 60, 90, and 120.

22

Asimov and Clarke and the Sacred Volume

It is interesting that the sacred volume is mars by factor of saturn to jupiter. The pillars of modern science fiction, Asimov and Clarke had as their pivots Saturn and Jupiter. In Clarke's Odyssey the strange events take place around Saturn in his book 2001, but around Jupiter in the screenplay and movie, 2001. In Asimov's Foundation and Earth, Jupiter and Saturn are clues to the location of Earth in the search for it by Golan Treviz and Janov Pelorat. Mars is the planet in our solar system that can be colonized or terraformed (altered to support life), Jupiter and Saturn are the most massive planets in the solar system. I have shown that mars is the cosmic fulcrum, the balancing point between Jupiter and Saturn, if they are placed on a teeter-totter. Mars would be the fulcrum when it is closest to Jupiter and Saturn and Jupiter are on opposite sides of the sun facing one another.

978-0-595-50477-0
0-595-50477-9